Food from Farmers

EGGS!

Life on a Chicken Farm

by Ruth Owen

WINDMILL BOOKS
New York

Published in 2012 by Windmill Books, an Imprint of Rosen Publishing
29 East 21st Street, New York, NY 10010

Editor for Ruby Tuesday Books Ltd: Mark J. Sachner
U.S. Editor: Julia Quinlan
Designer: Emma Randall
Consultant: Logan Peterman, Laughing Sprout Family Farm

Photo Credits: Cover, 1, 4–5, 6–7, 8–9, 10–11, 12–13, 14–15, 16, 17 (top), 17 (center), 18 (top), 18 (bottom), 19 (top), 20, 21(top), 22–23, 24–25, 26–27, 28–29, 30–31 © Shutterstock; 17 (bottom), 18 (center), 19 (center) © Wikipedia (Creative Commons public domain); 21 (bottom) © istockphoto.

Library of Congress Cataloging-in-Publication Data

Owen, Ruth, 1967–
Eggs! : life on a chicken farm / by Ruth Owen.
p. cm. — (Food from farmers)
Includes index.
ISBN 978-1-61533-529-9 (library binding) — ISBN 978-1-61533-536-7 (pbk.) —
ISBN 978-1-61533-537-4 (6-pack)
1. Eggs—Juvenile literature. 2. Egg gathering—Juvenile literature.
3. Chickens—Juvenile literature. I. Title.
SF490.3.O94 2012
636.5'142—dc23

2011022782

Manufactured in the United States of America

CPSIA Compliance Information: Batch #BOW2102WM: For Further Information contact Windmill Books, New York, New York at 1-866-478-0556

CONTENTS

WELCOME TO MY FARM!

Hi! My name is Lauren. I am almost 12 years old. I live on a farm with my mom, dad, and two little brothers. Our farm is a chicken farm. We produce eggs to sell.

Only female chickens lay eggs. Female chickens are called hens. We have about 1,000 hens on the farm.

My dad works on the farm every day of the year, even when it's a holiday! Mom is a teacher, but she has to help out on the farm, too.

This is Ginger, my pet hen. Last year, Ginger won first prize in a chicken show. It was like a beauty contest for chickens!

GINGER'S EGG FACTS

- **There are around 305 million hens in North America. They lay about 66 billion eggs each year!**

LET'S LOOK AROUND THE FARM

This is a map of our farm. We live in the farmhouse. We own 30 **acres** of land.

GINGER'S FARM FACTS

- **An acre is a measurement used to measure farms.**
- **An acre is just a little smaller than a football field.**
- **An acre equals 0.4 hectare.**

During the spring and summer, we raise turkeys to sell at Thanksgiving.

Turkeys

On many chicken farms, the hens spend all their time in a barn. Our farm is a **free-range** chicken farm. This means the hens get to live outside in the **pasture**.

This is a field of sunflowers. The hens like the shade under the flowers.

When the sunflowers are fully grown, Dad will chop them down. Then the hens will eat the sunflower seeds.

MEET THE HENS

We keep many different types of hens on our farm.

Rhode Island red hen

Araucana hen

Plymouth rock hen

When they are young, female chickens are called pullets. A pullet starts to lay eggs when she is about six months old.

At first the pullet's eggs are small. As she gets older, the eggs reach full size. Now she is a hen!

A hen lays about two eggs every three days. Some hens lay an egg every day, though!

GINGER'S HEN FACTS

- It takes a hen up to 26 hours to make an egg inside her body. She lays the egg, and then, after about 30 minutes, she starts making another egg all over again!

These hens are inside a cozy box inside one of the nesting houses.

Each hen on our farm lays about 240 eggs in a year. That's 240,000 eggs each year among all our girls!

HENS AND ROOSTERS

We all know that baby chickens, called chicks, come from eggs. So why do some eggs turn into chicks?

A male chicken, or rooster, is needed to make a chick.
The rooster **mates** with the hen.
Now, the hen's egg is **fertilized**.
This means a chick can grow inside the egg.

The hen sits on her egg.
This keeps the egg warm and helps the chick grow.
After 21 days, the chick **hatches**!

GINGER'S EGG FACTS

- **A hen turns her eggs in the nest about 50 times each day so the insides don't stick to the shell.**

Our farm produces eggs, not chickens, so we don't need roosters. Dad lets me keep some pet hens and one rooster, though.

This is Charlie, my pet rooster.

Sometimes Charlie and my girls have chicks!

CHICK TIME!

When we need some new hens, we buy female chicks from a place called a hatchery.

At the hatchery, thousands of hens and lots of roosters produce fertilized eggs.

The eggs are put in large warm boxes called **incubators**. The warmth makes the chicks hatch. Then they are ready to go to chicken farms.

Dad collects the chicks in the truck. The chicks are just one day old! They arrive in cardboard boxes.

We make a pen in the barn for the chicks.
My brothers and I give them some feed.

The chicks might be thirsty after their trip. We dip each chick's beak in the water bowl to show it where to find water.

The chicks can warm up under a light bulb in here.

GINGER'S CHICK FACTS

- **Chicks can't stay warm on their own. Using a light bulb, farmers make the chicks a warm place to take shelter.**

WHAT DO THE HENS EAT?

Our free-range hens eat grass, bugs and worms out in the pasture.

GINGER'S HEN FACTS

- **Chickens like to scratch at the ground to dig up insects and worms to eat!**

We give the hens a special chicken food and corn, too. Dad drives around the pasture in the truck and fills up the feeders.

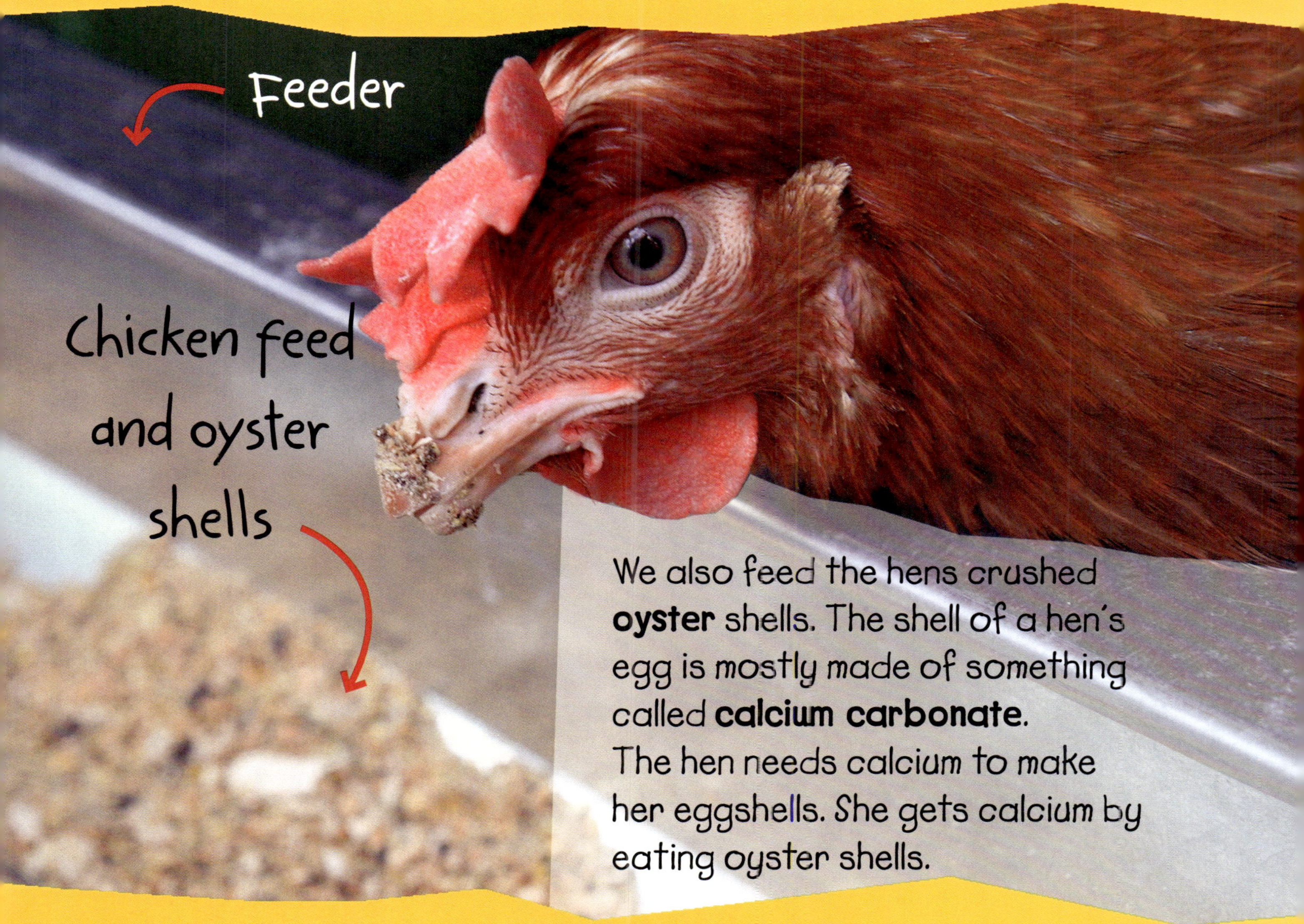

We also feed the hens crushed **oyster** shells. The shell of a hen's egg is mostly made of something called **calcium carbonate**. The hen needs calcium to make her eggshells. She gets calcium by eating oyster shells.

Chickens don't have teeth to crunch up their food, so they eat grit. The grit gathers inside the chicken's body in a body part called a gizzard. Inside the gizzard, the chicken's body uses the grit to grind up the food!

COLLECTING THE EGGS

Hens like to lay their eggs in a dark, quiet place.

Dad has built our hens' nesting houses out in the pasture. Inside the nesting houses, there are nesting boxes. The hens lay their eggs in the nesting boxes.

Dad drives out to the nesting houses in the truck. He collects the eggs from the nesting boxes. He collects eggs every day, in the morning and in the afternoon.

Some hens lay white eggs, and others lay brown eggs. Hens with red ear lobes, like Ginger, lay brown eggs. Hens with white ear lobes lay white eggs.

Brown, white, or blue eggs taste exactly the same.

GINGER'S EGG FACTS

- **Araucana chickens lay bluish-green eggs! Araucanas are nicknamed the "Easter egg chicken."**

PROCESSING THE EGGS

After Dad collects the eggs, it is time to get them ready for sale. In the barn, we have a machine that **processes** the eggs.

Dad carefully lays the eggs on the machine's conveyor belt.

First, the eggs are **candled**. This means that the machine shines a light through the eggs. Candling allows Dad to see inside each egg and make sure it has a good, strong yolk.

GINGER'S EGG FACTS

- **Checking inside an egg is called candling because years ago, people held candles behind eggs to check inside them.**

Then, the eggs are washed with brushes. It's like a car wash for eggs!

Finally, the machine weighs each egg. This tells Dad if the egg should be sold as a small, medium, large, or jumbo egg.

WHAT HAPPENS AT AN EGG FACTORY?

Some farms send their eggs to a large egg factory, or processing plant.

The machines at the plant are much bigger than Dad's!

GINGER'S EGG FACTS

- **A large, high-tech egg processing plant can process up to two million eggs each day!**

At the plant, machines wash the eggs in hot water. Cameras then photograph the eggs. A computer quickly checks the photos. It looks for any eggs that still have dirt spots and must be washed again.

Next, the eggs go through the crack detector. A machine gently taps each egg and listens to the sound it makes. A cracked egg will make a different sound than an egg with no cracks.

Then the eggs are candled and weighed. To make sure the eggs are sized exactly right, a machine weighs each egg 60 times in less than a second!

Finally, the machines put the eggs into cartons. Then the cartons are stacked in cardboard boxes by robots!

EGGS FOR SALE!

Some chicken farms send their eggs to a processing plant. Then the plant sells the eggs to large supermarkets.

We sell our eggs to local people and businesses.

Local stores, restaurants, and bakeries email Dad with their egg orders.

Dad packs the orders in the truck and delivers them to our customers.

This is Phil. His bakery makes great muffins!

We also sell eggs from a stand at the gate to our farm.

On Saturday, we have a stand at a farmers' market.

At the market, farmers sell the things they produce. You can buy fruit, vegetables, meat, cheese, bread, and honey.

GINGER'S EGG FACTS

- **Eggs are normally sold twelve in a carton. This is called a dozen.**

Some people make and sell crafts. My knitted hat was made using wool from sheep that live near our farm!

A DAY IN THE LIFE OF A FARM

A chicken farmer works very hard!

7:00 a.m.
Dad is out in the pasture. He collects 200 eggs from the nesting houses!

9:00 a.m.
Dad fills up the hens' feeders with food.

11:00 a.m.
The egg machine and Dad process the 200 eggs.

1:00 p.m.
Dad drives into town. He delivers eggs to five stores.

GINGER'S FARM FACTS

- Chicken farms often have electric fences. The fence wires give off an electric shock if they are touched. Electric fences keep out animals, such as foxes, that want to catch chickens.

3:00 p.m.
Some fences around the pastures need fixing.

5:30 p.m.
Mom and Dad collect 350 eggs from the nesting houses.

7:00 p.m.
Dad makes us omelettes for supper with fresh eggs!

8:00 p.m.
Mom checks emails from our customers who want to order eggs.

FREE-RANGE HENS ARE HAPPY HENS!

We are proud that our farm is a free-range chicken farm.

My brother feeds some hens and our pet goat.

On some chicken farms, the hens spend all their lives inside a barn. These farms are often called factory farms.

GINGER'S HEN FACTS

- **Factory farm hens lay their eggs onto a conveyor belt. They cannot go into a quiet nesting box.**

Factory farm hens are kept in small cages. The hens cannot walk around or flap their wings. They don't get the chance to eat grass or scratch for insects.

If a factory farm hen stops laying for a while, the farm gets rid of her. Sometimes Dad buys these hens. We give them new lives on our farm!

Rescue hen

The hens' feathers have rubbed off on the bars of the cages. The hens can't walk well. I take care of the rescue hens. They soon get better and are ready to be free-range hens!

WE LOVE EGGS!

Eggs taste great, and they contain 13 important **nutrients**. Nutrients are all the good things that a body needs to keep working.

Eggs contain **protein**. This nutrient keeps your muscles healthy and strong.

Protein helps fill you up, too. Having eggs for breakfast keeps you full until lunch. This means you can do your schoolwork. You won't be thinking about your stomach rumbling with hunger!

Eggs have **vitamin** A, which keeps your eyes and skin healthy. They also have vitamin D for strong bones and teeth.

These are my favorite ways to eat eggs.

French toast with fruit for breakfast.

An egg salad sandwich for lunch.

Thank you to chicken farmers and hens everywhere!

GINGER'S AMAZING EGG FACTS

- Eggs are used in baking because they help hold all the other ingredients, such as flour and sugar, together.

GLOSSARY

acre (AY-kur)
A unit of measurement used for measuring land, especially on farms.

calcium carbonate
(KAL-see-um KAR-buh-nayt)
A substance found in rocks. It is also one of the main ingredients of snail shells, seashells, and the shells of bird eggs.

candled (KAN-dld)
When an egg is held up to a light so that the inside of the egg can be checked to make sure it is healthy and has a strong yolk.

fertilized (FUR-tuh-lyzd)
Eggs that can grow into chicks.

free-range (FREE RAYNJ)
Where animals are allowed to roam freely and live outside.

hatches (HATCH-ez) Breaks out of an egg.

incubator (ING-kyuh-bay-ter)
A heated container used for keeping eggs, or sometimes baby animals or human babies, warm. An incubator can be as small as a shoebox or as big as a room.

mate (MAYT)
When a male and a female animal get together to produce young.

nutrients (NOO-tree-ents)
Substances that the body needs to help it live and grow. Foods contain nutrients such as vitamins and protein.

oyster (OY-stur)
A small ocean animal with a shell.

pasture (PAS-chur)
A field where farm or pet animals, such as chickens, sheep, or horses, can eat grass.

processes (PRAH-ses-ez)
Carries out a series of actions to prepare or change something.

protein (PROH-teen)
A nutrient needed by the body. It is found in foods such as eggs, milk, meat, and fish.

vitamin (VY-tuh-min)
A substance found in foods that is needed by the body for health and growth.

WEB SITES

For Web resources related to the subject of this book, go to: www.windmillbooks.com/weblinks and select this book's title.

READ MORE

Dickman, Nancy. *A Chicken's Life*. Portsmouth, NH: Heinemann Educational Books, 2010.

Mercer, Abbie. *Chickens on a Farm*. New York: PowerKids Press, 2010.

Metz, Lorijo. *Chickens: Hens, Roosters, and Chicks*. New York: PowerKids Press, 2011.

INDEX